Ultimate Dot to Dot Cats
Extreme Stress Relief and Relaxing Challenges
Puzzles From 150-411 Dots

By Laura's Dot to Dot Therapy

Copyright © 2018

All rights reserved. No part of this publication may be reproduced, distributed, or transmitted in any form or by any means, including photocopying, recording, or other electronic or mechanical methods, without the prior written permission of the publisher

How To Use This Book

Hi! We're so glad you're a lover of puzzles and dot connecting- we are too!

Connecting the dots in this book is simple- just relax and follow the numbers in consecutive order, drawing a straight line between each one. Dot 1 will connect to dot 2 and so on and so forth until there are no more dots to connect. There's always another dot and you'll always find it. Connect every dot to discover the beautiful images they create.

In case you get lost or can't find a dot, never stress- there's an answer key at the back of the book that will show you exactly where each dot connects to the next. If you want to color your images, we encourage you to do so! Feel free to try all different colors and coloring mediums for your images!

If you find any errors or omissions in this book, email us at Laurasdottodot@gmail.com and please let us know! We want you to have the best dot to dot experience!

Image 1

Image 2

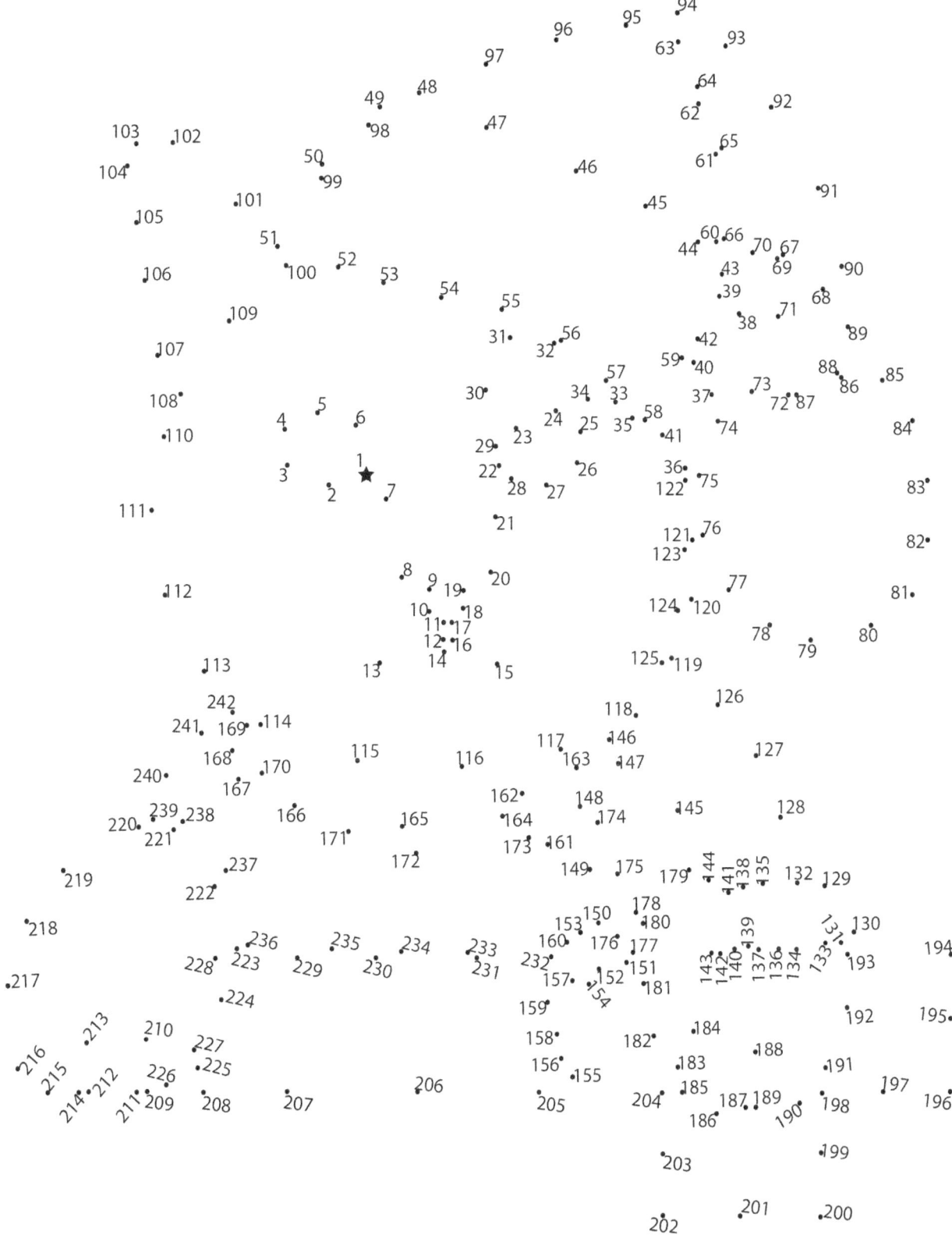

Image 3

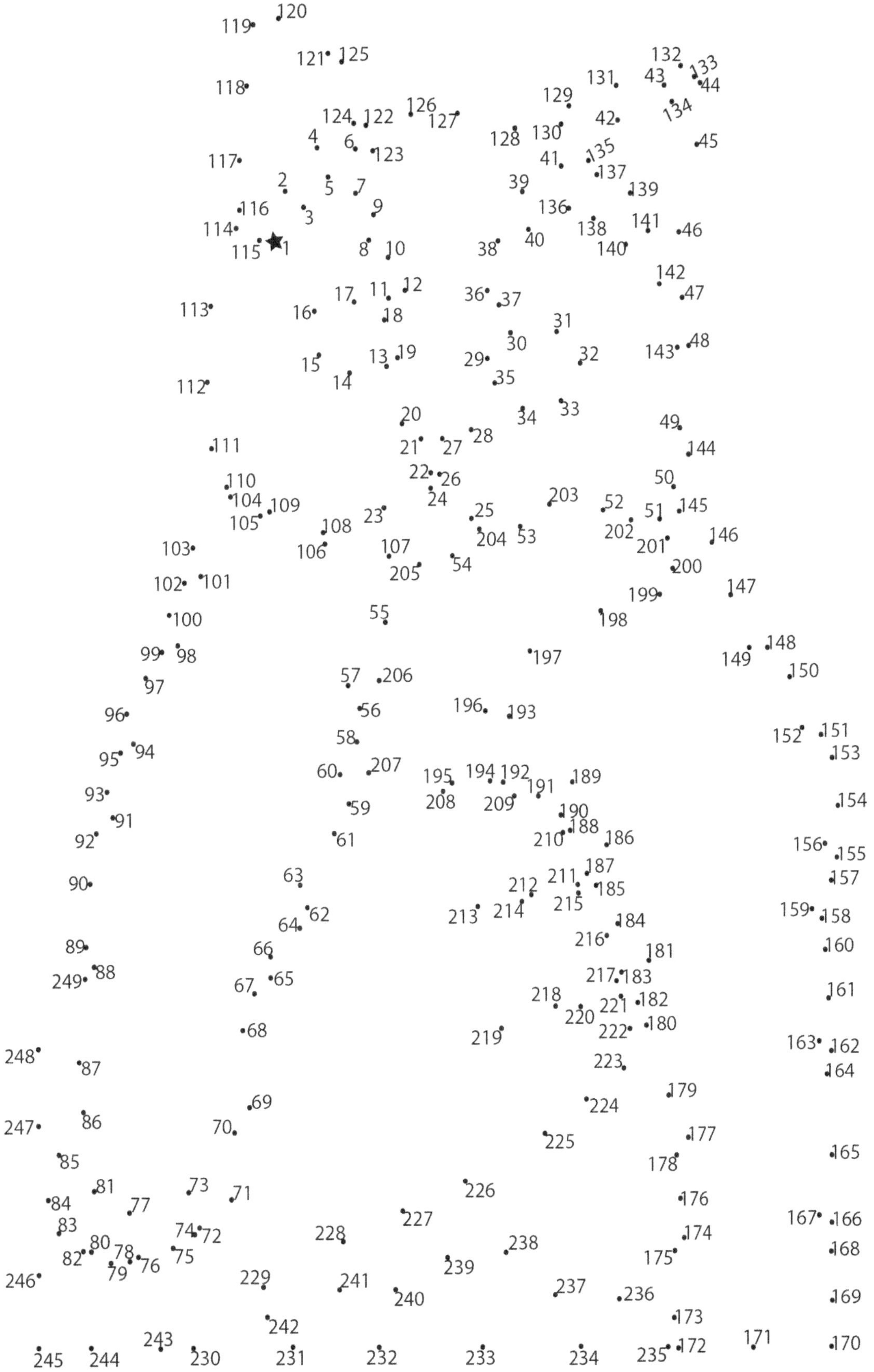

Image 4

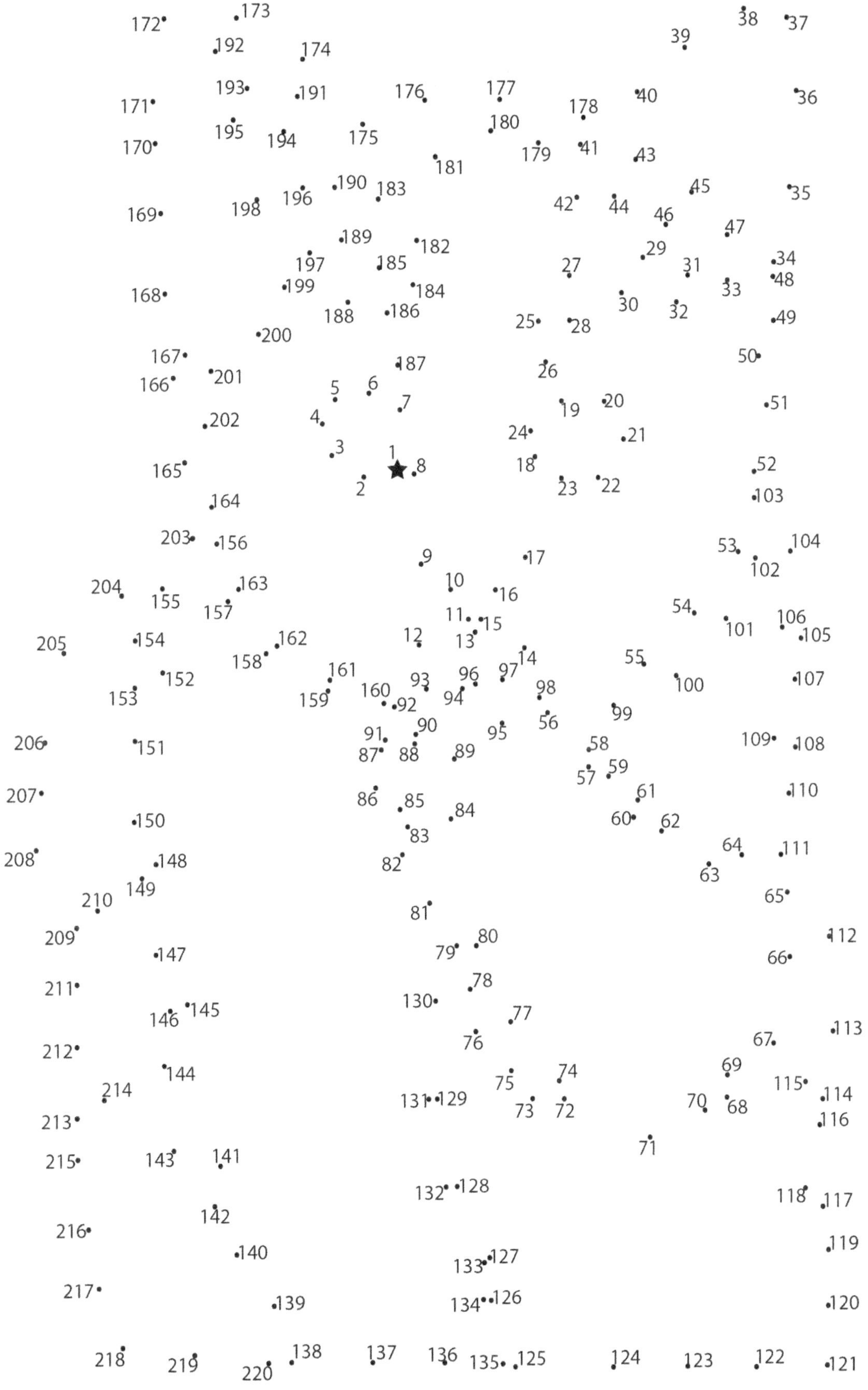

Image 5

Image 6

Image 7

Image 8

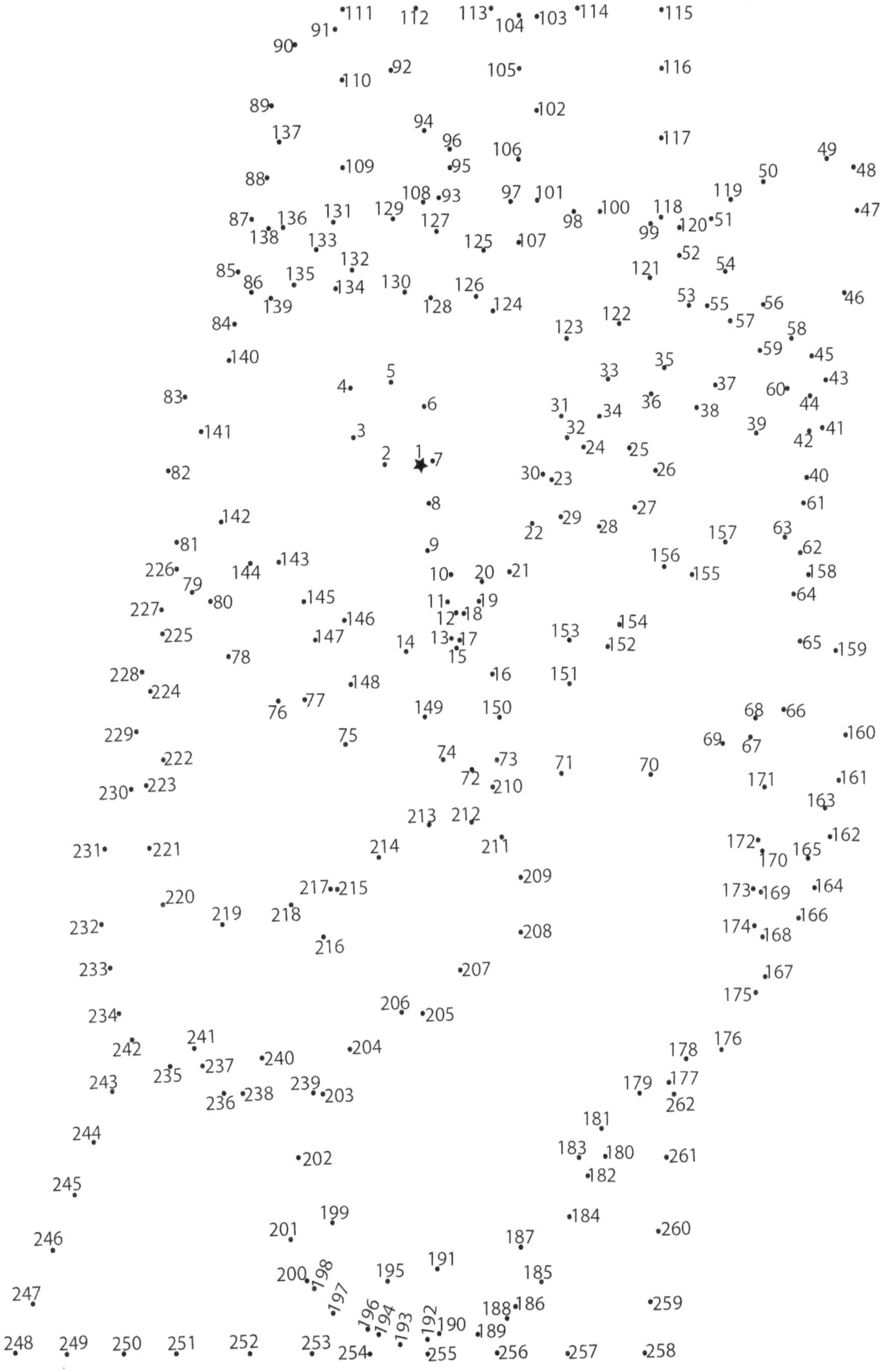

Image 9

Image 10

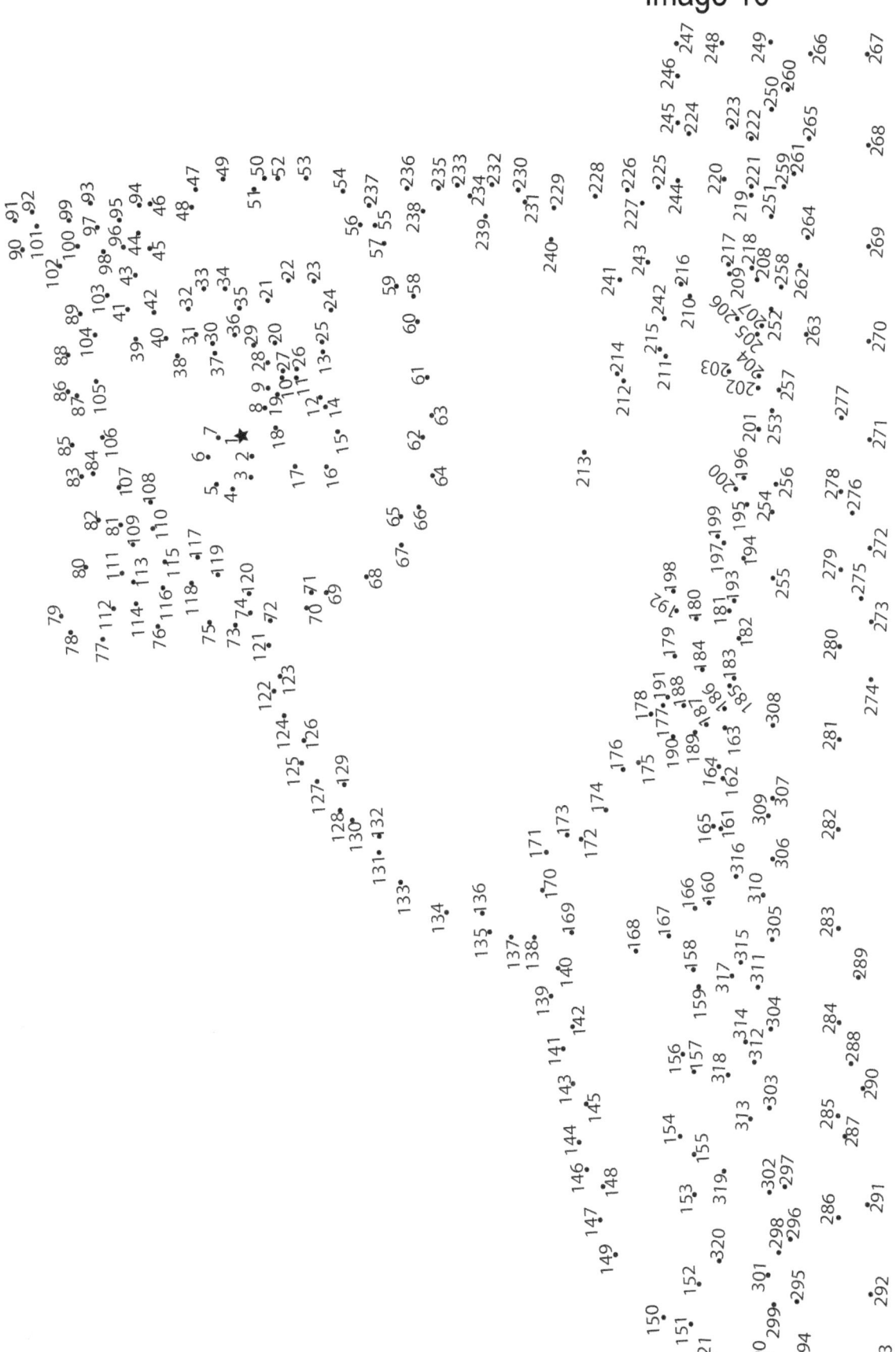

Image 11

Image 12

Image 13

Image 14

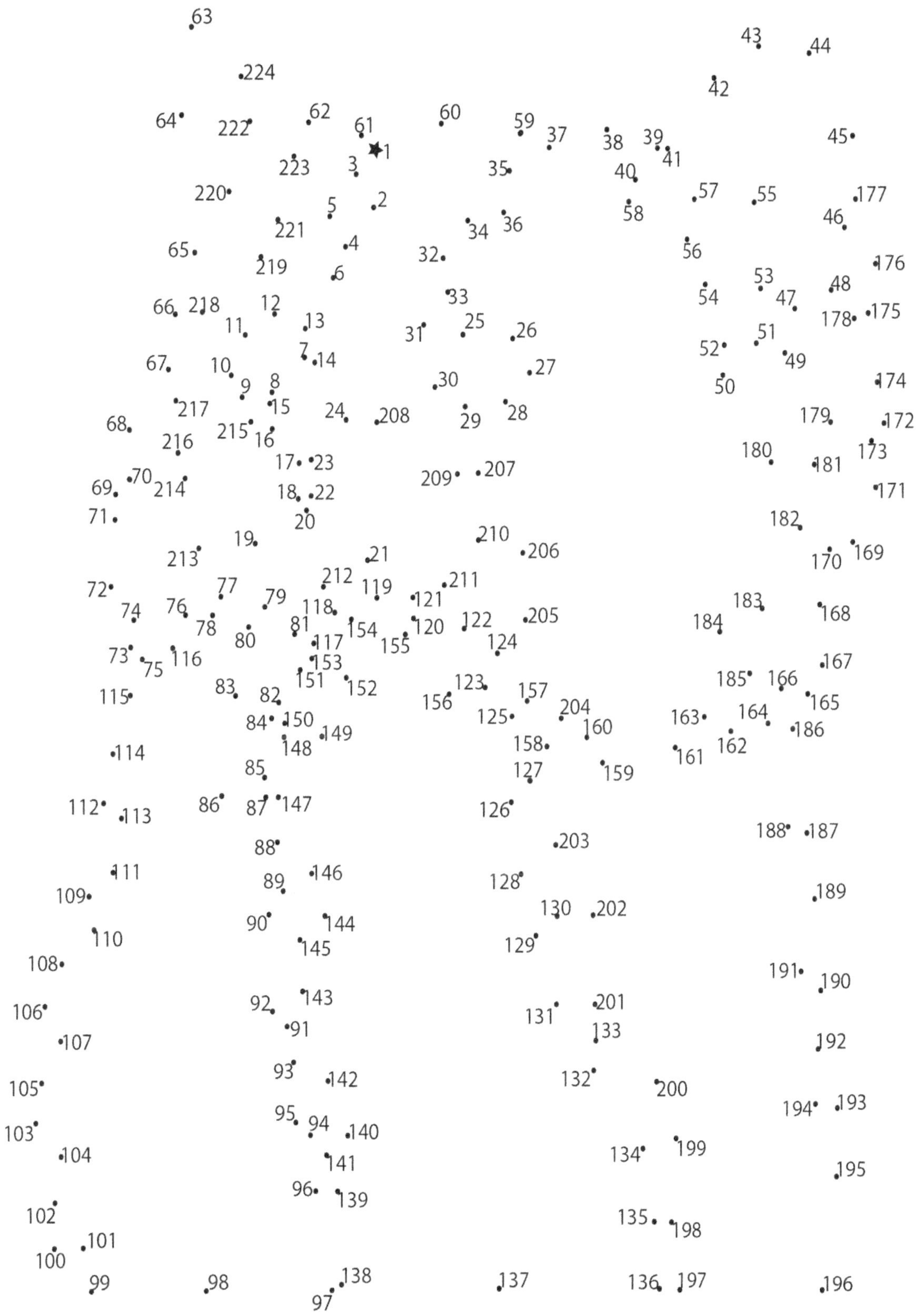

Image 15

Image 16

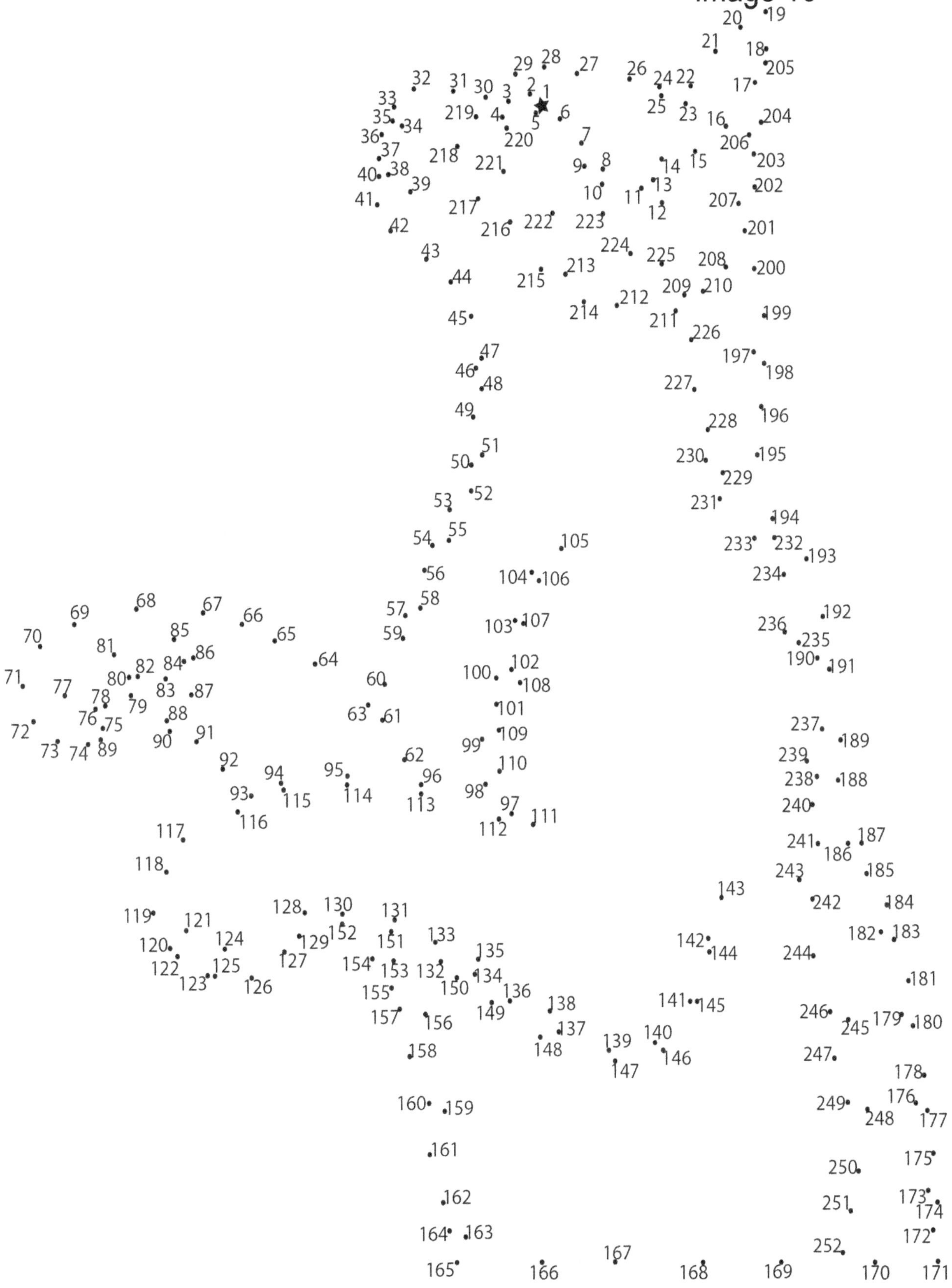

Image 17

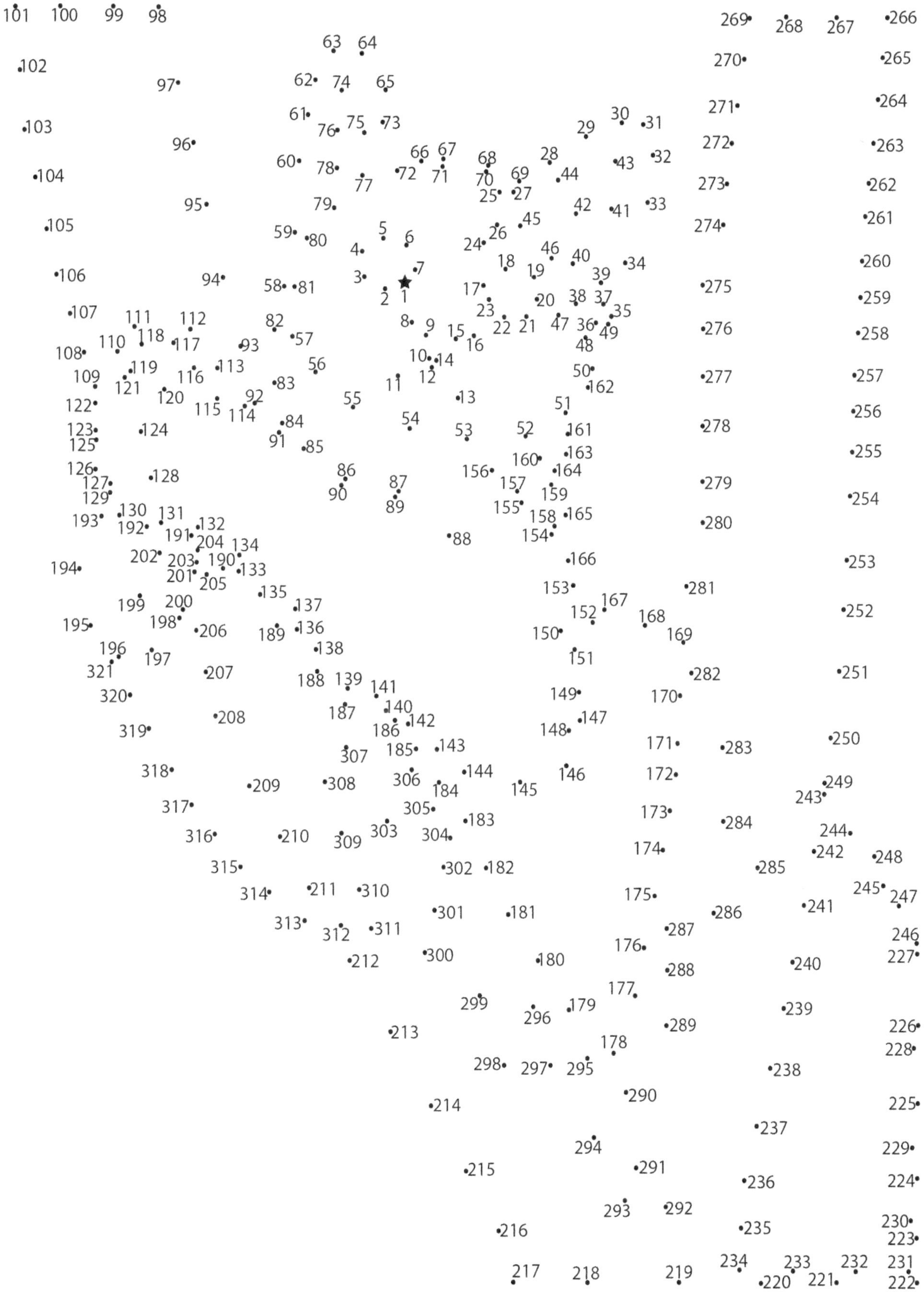

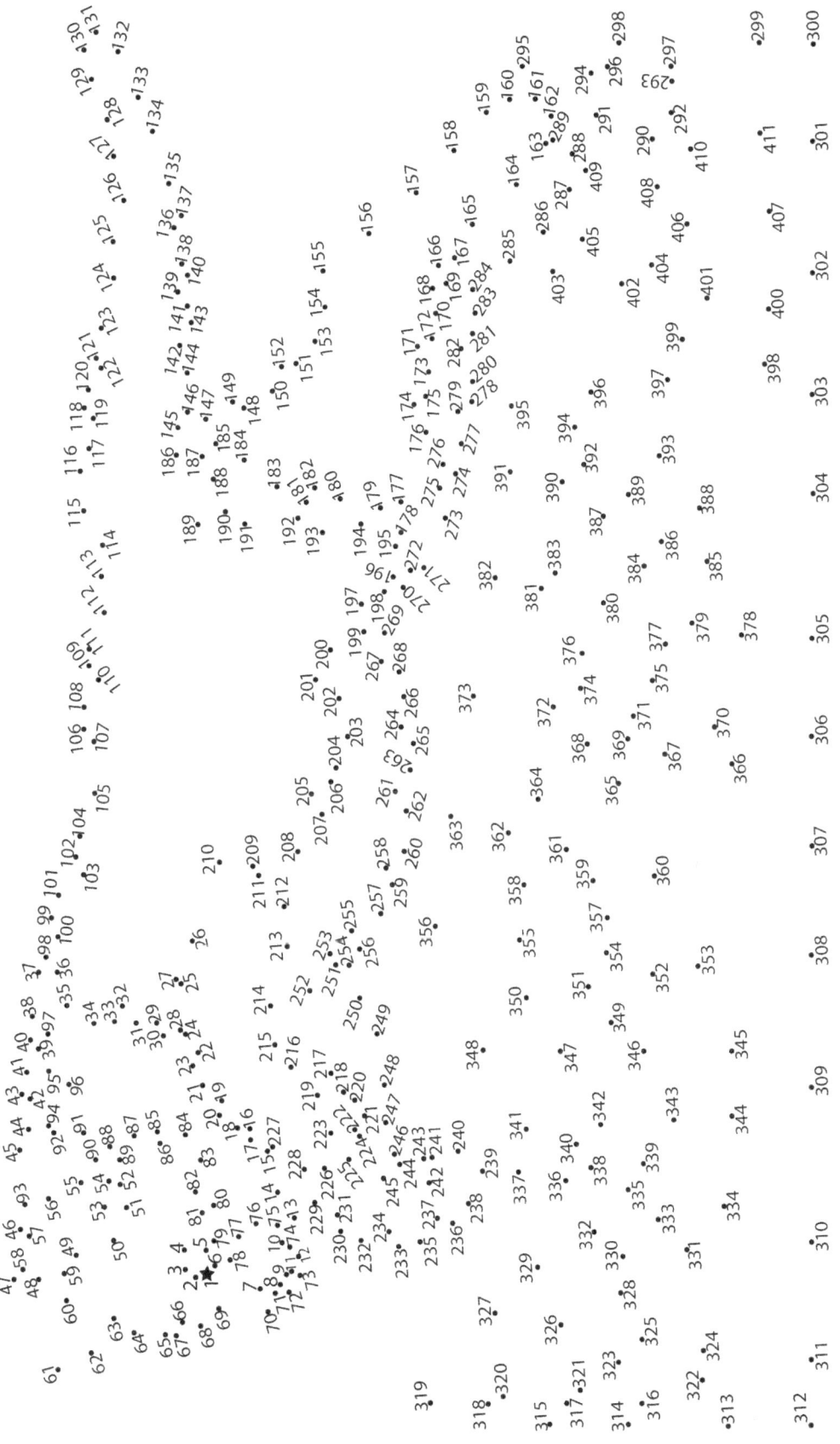

Image 18

Image 19

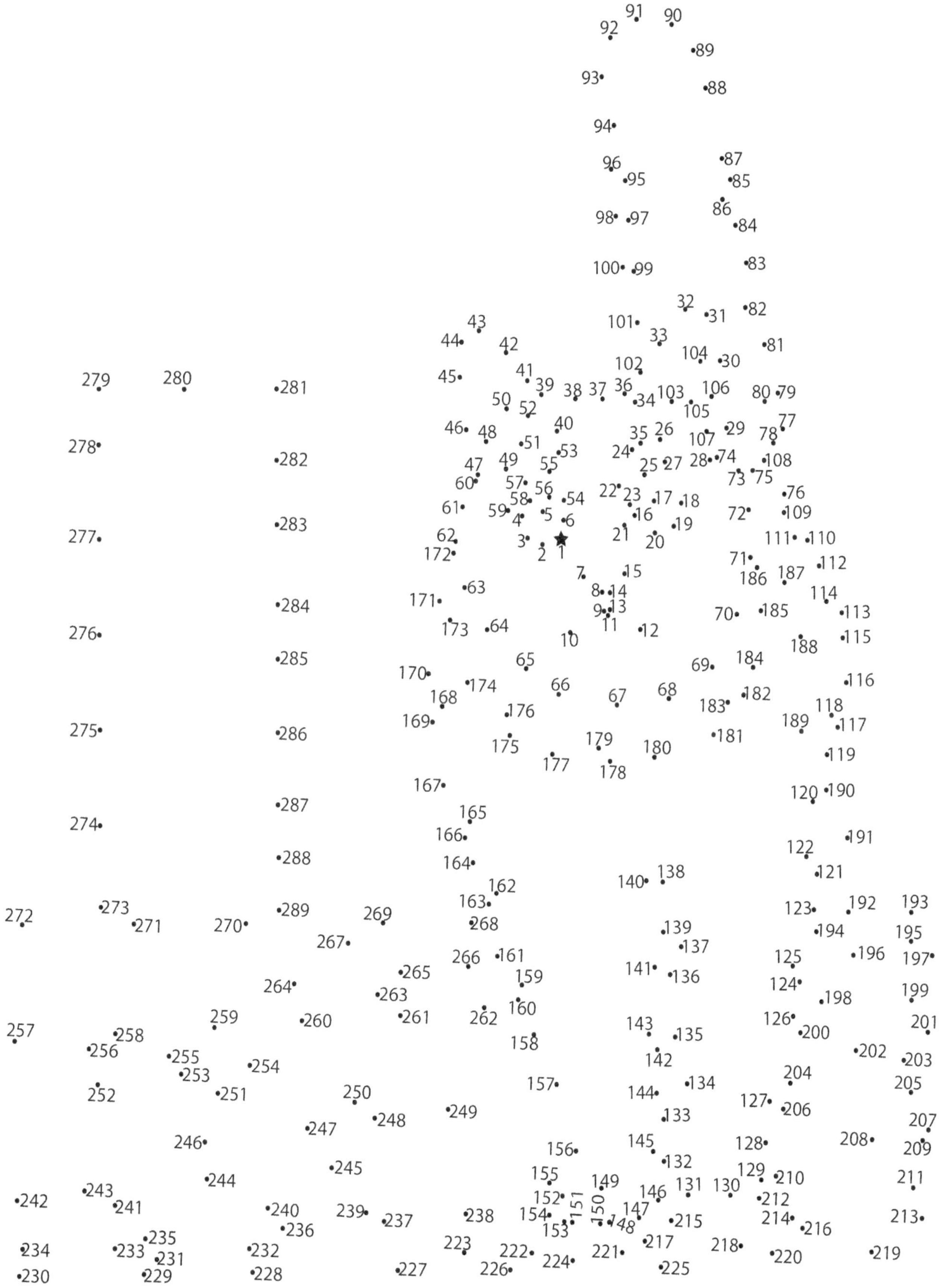

Image 20

Enjoy bonus images from some of our other fun dot-to-dot books

Find all of our books on Amazon

Easy to Read Large Print Dot to Dot
Beautiful Landscapes
Puzzles from 150 to 760 Dots

Antique Cars and Vintage Cars
Large Print Dot-to-Dot Book for Adults

Famous Faces
Big Book of Extreme Dot to Dot

Mythical Mermaid
Dot-to-Dot Book for Adults
Puzzles from 150 to 750 Dots

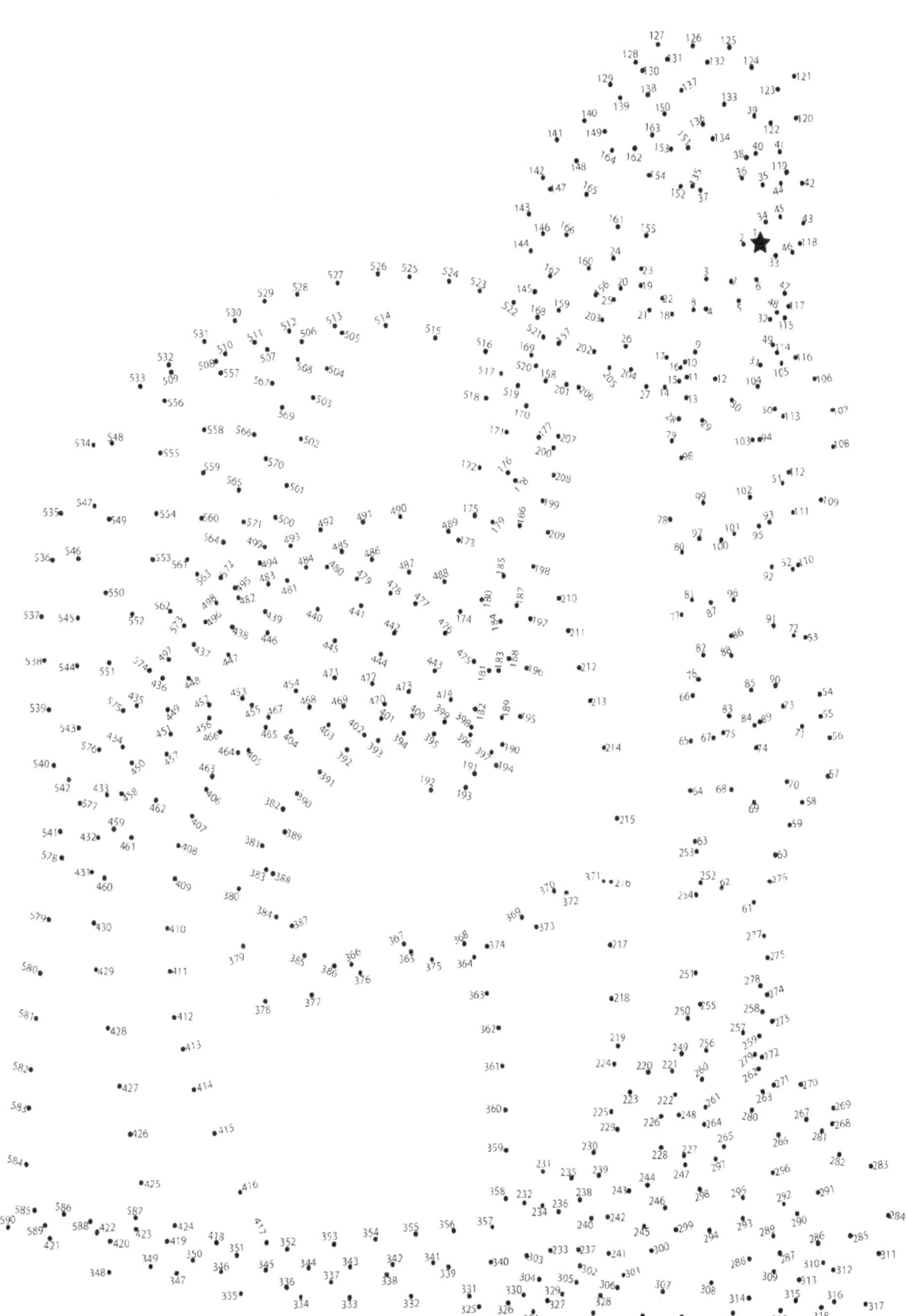

Unicorn, Dragons, and Magical Creatures Dot-to-Dot Puzzles from 462 to 956 Dots

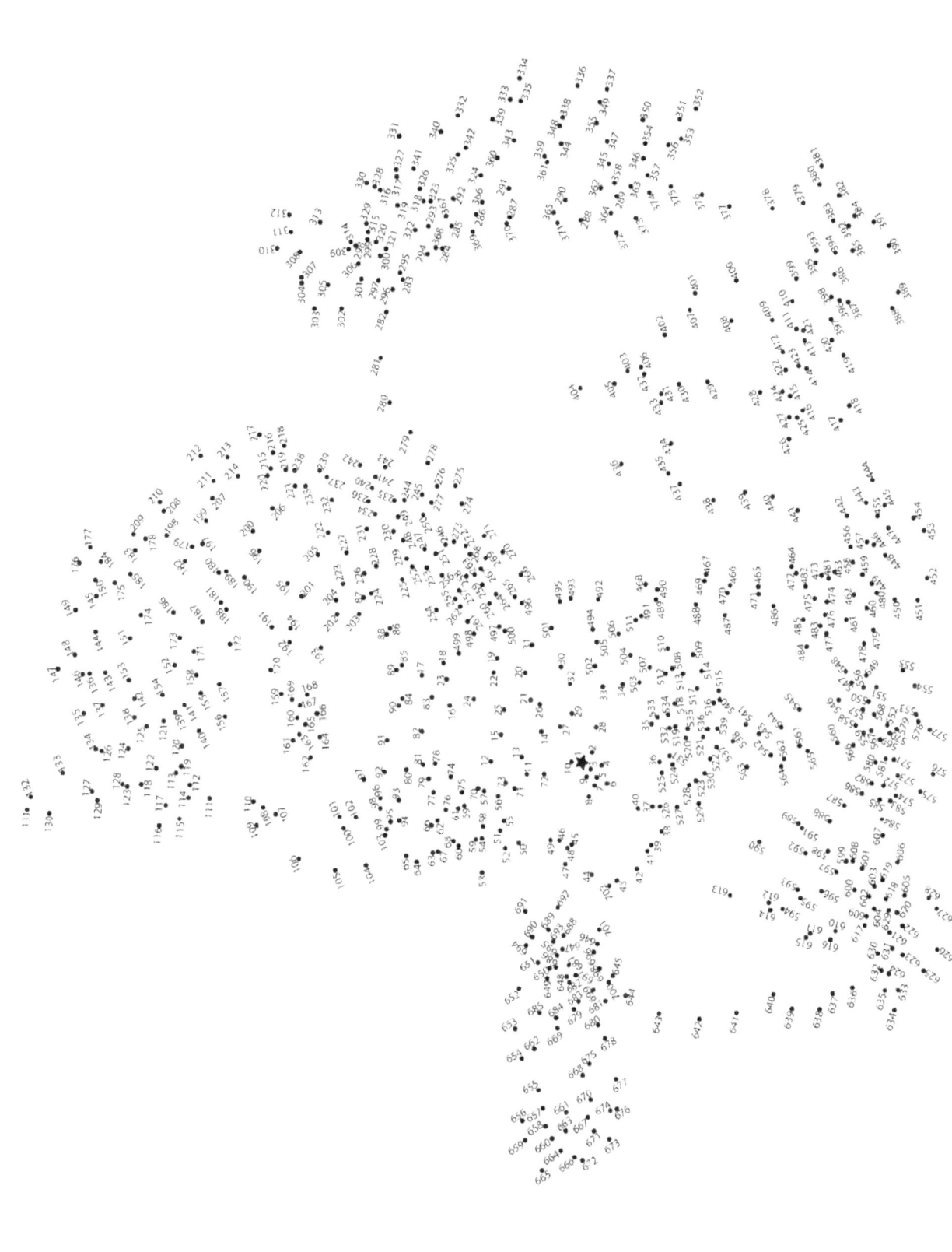

Answer Key

Follow along with the
page numbers from top left
to bottom right

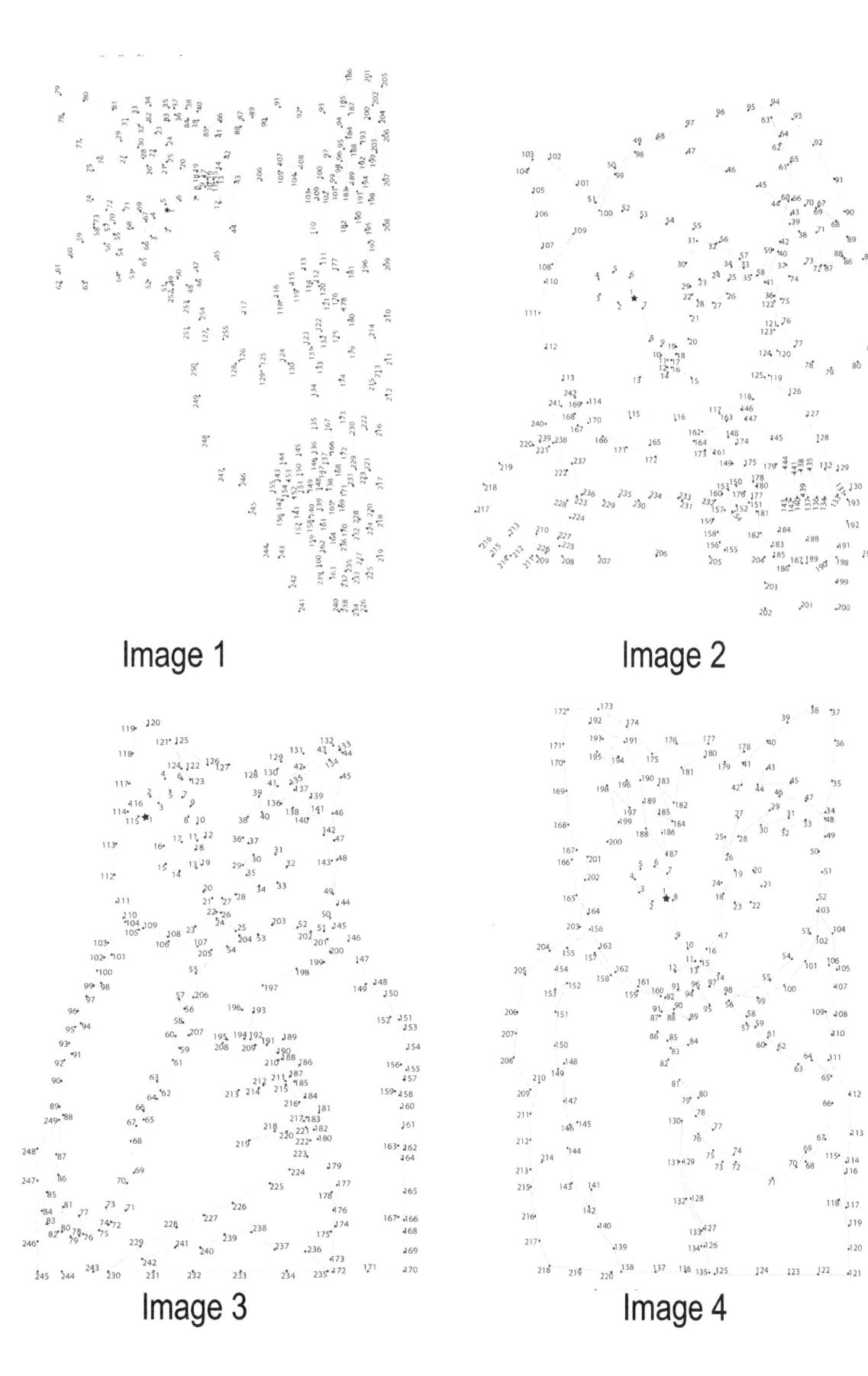

Image 1

Image 2

Image 3

Image 4

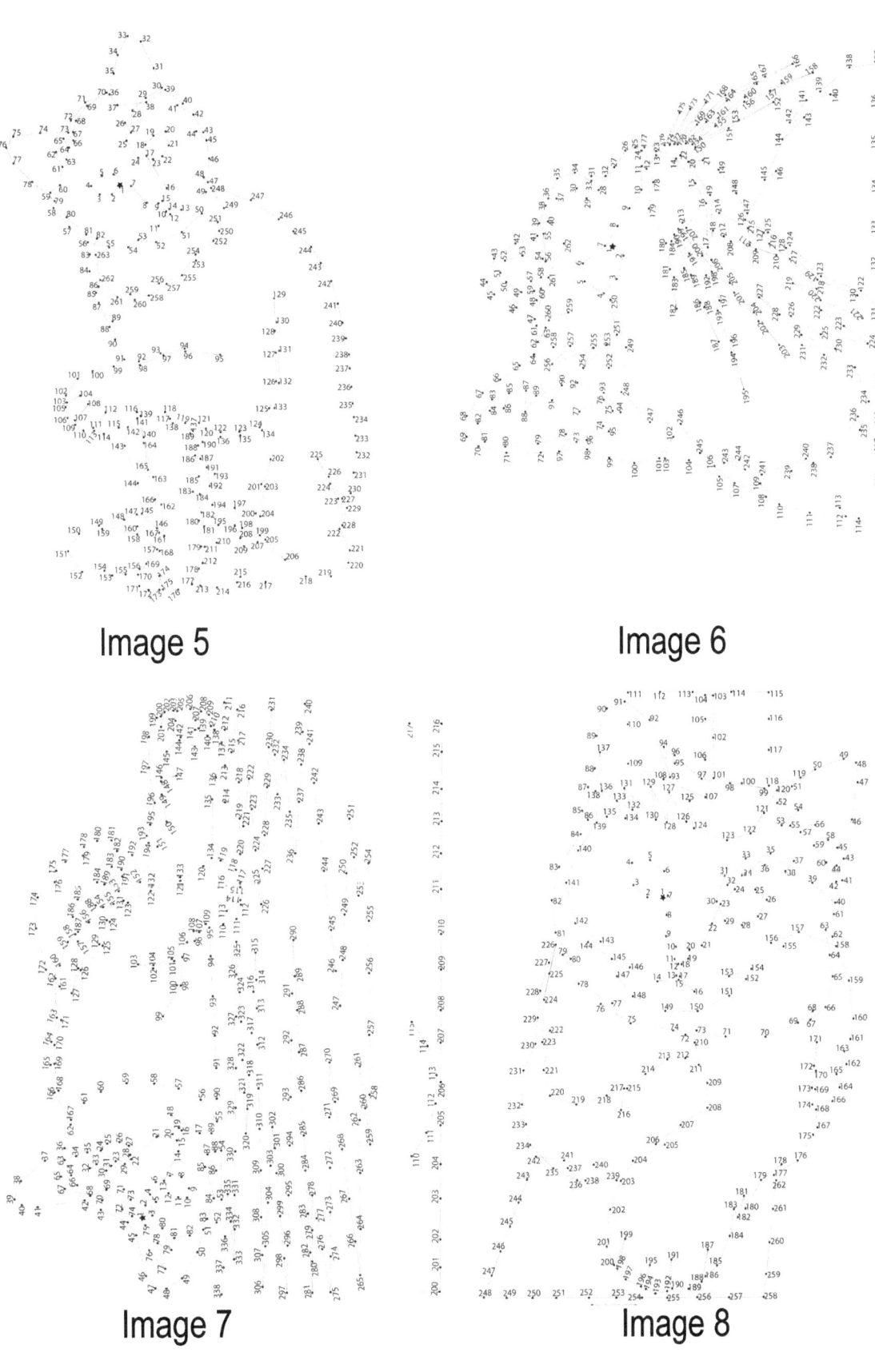

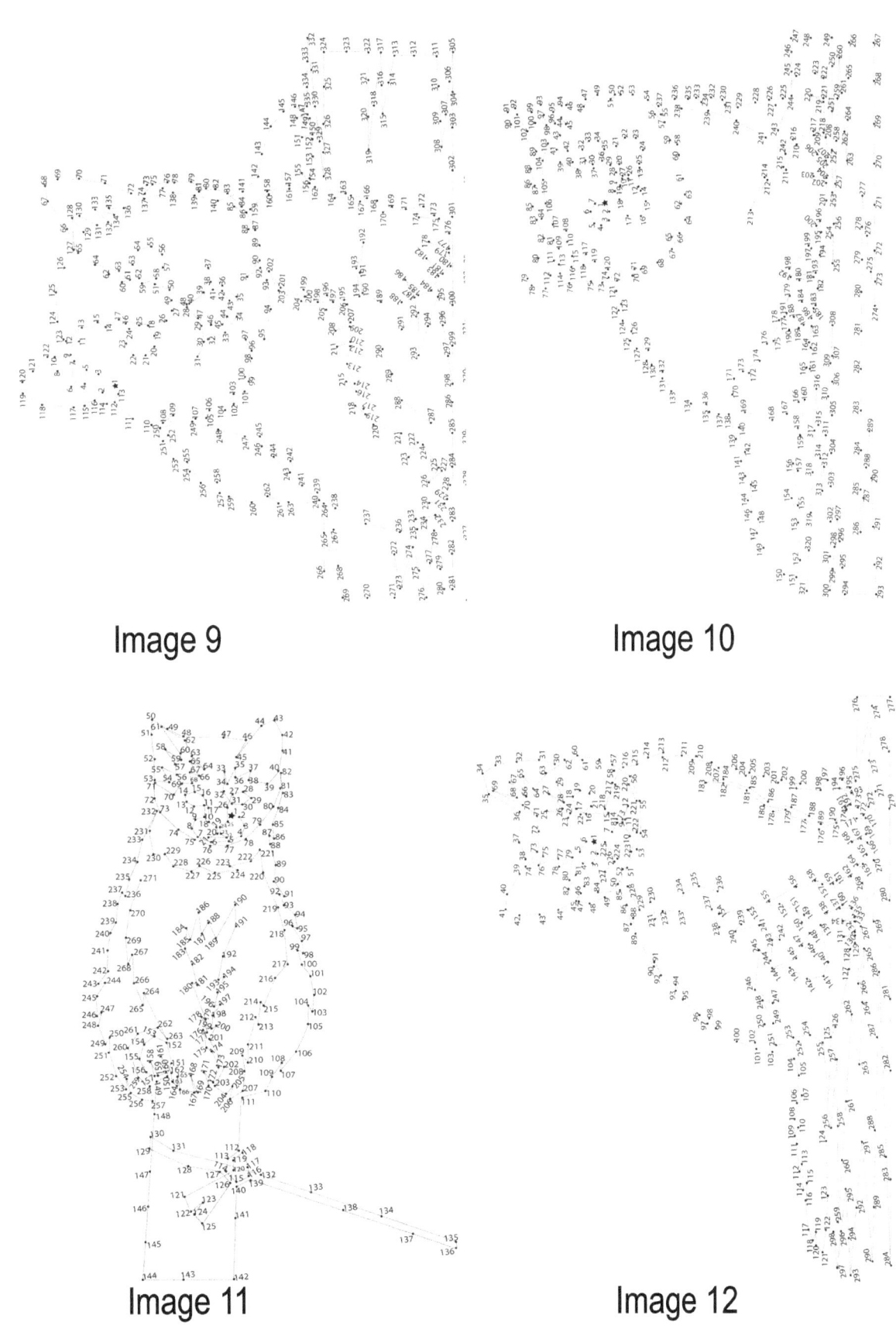

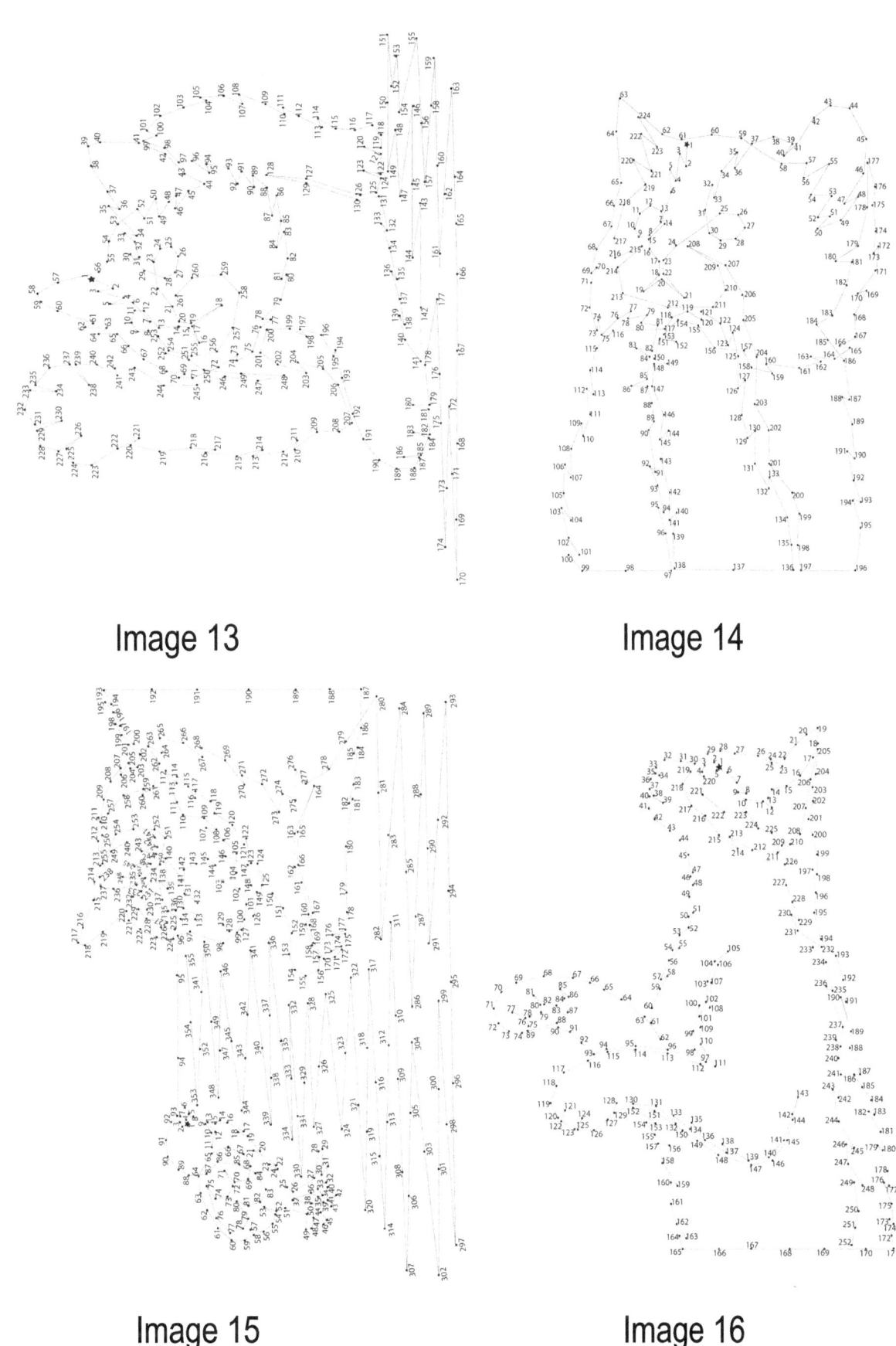

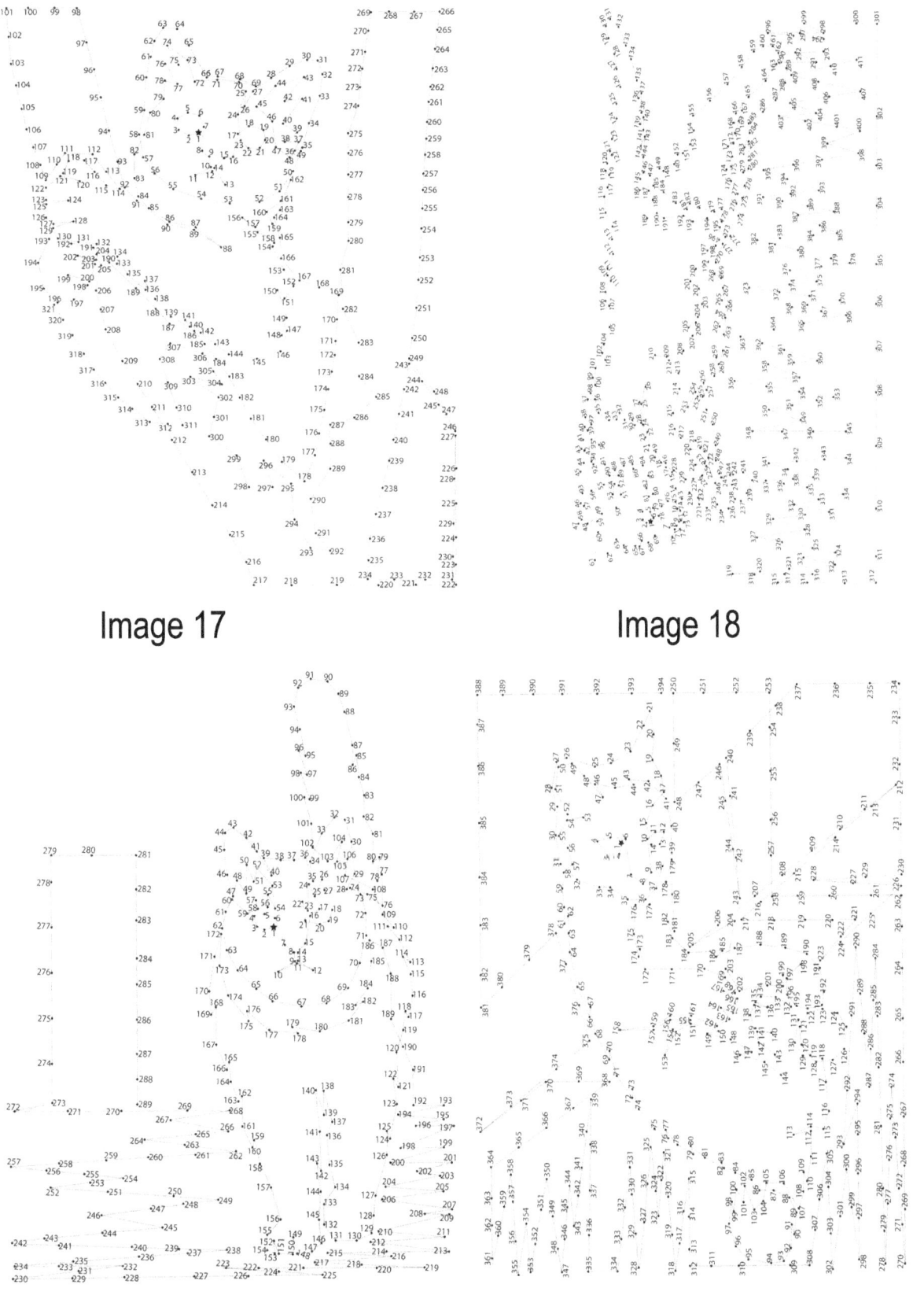

www.ingramcontent.com/pod-product-compliance
Lightning Source LLC
Chambersburg PA
CBHW082015230526

45468CB00022B/2326